BEYO

SURVE

MAD

An original Book by

Jim Crume P.L.S., M.S., CFedS, RP

Co-Authors
Cindy Crume
Bridget Crume, RP
Aaron Michalenko PLS, RP
Richard Delgado, CP
Tony Slater, PLS

KINDLE - PRINTED EDITIONS

*****

PUBLISHED BY:

Jim Crume P.L.S., M.S., CFedS, RP

BEYOND THE BASICS

Copyright 2018 © by Jim Crume P.L.S., M.S., CFedS, RP

All Rights Reserved

First publication: January, 2018

Cover photo: I-10 and Valencia Road T.I.
Tucson, Arizona

TERMS AND CONDITIONS

The content of the pages of this book is for your general information and use only. It is subject to change without notice.

Neither we nor any third parties provide any warranty or guarantee as to the accuracy, timeliness, performance, completeness or suitability of the information and materials found or offered in this book for any particular purpose. You acknowledge that such information and materials may contain inaccuracies or errors and we expressly exclude liability for any such inaccuracies or errors to the fullest extent permitted by law.

Your use of any information or materials in this book is entirely at your own risk, for which we shall not be liable. It shall be your own responsibility to ensure that any products, services or information available in this book meet your specific requirements.

This book is covered by the Amazon Publishing Terms and Conditions.

This book may not be further reproduced or circulated in any form, including email. Any reproduction or editing by any means mechanical or electronic without the explicit written permission of Jim Crume is expressly prohibited.

TABLE OF CONTENTS

INTRODUCTION 4
Beyond the Basics 8
Part 107 Rules and Regulations 9
National Airspace System 10
Waivers and Airspace Authorization 13
Local Drone Laws and Ordinances 16
NOTAM'S 16
METAR's 17
Coordinated Universal Time (UTC) 22
Urban Environment 23
Rural Environment 26
Drones and the Public 26
Flight Logs 28
Sample Missions 29
Checklist 35
Conclusion 35
Other books in this series 36
Other books by this author 37
ABOUT THE AUTHOR 38

INTRODUCTION

Straight forward Step-by-Step instructions.

This book is just one part in a series of digital and paperback books on Survey Mapping Made Simple. The subject matter in this book will utilize the methods for survey mapping with a drone (a.k.a. sUAS, UAS, UAV)

For a list of books in this series, please visit:

http://www.cc4w.net/drone.html

Prerequisites for this book series:

Surveying knowledge and experience is required to understand the principles and applications for survey mapping that will be presented throughout this book series.

A Part 107 FAA Remote Pilot Certification is required to utilize the methods shown in this book series.

If you have not studied, tested and received your Part 107 FAA Remote Pilot Certification then use the following information to help you in achieving that objective:

Start by looking at the FAA website for Unmanned Aircraft Systems (UAS).

https://www.faa.gov/uas/

There are plenty of online resources for learning the required information in preparation for the exam. It can be a daunting task to sort through all that is available.

I personally had great success with Remote Pilot 101.

BEYOND THE BASICS

http://remotepilot101.com/

The instructor did an excellent job in covering the items needed to pass the FAA exam such as:
- Rules and Regulations
- Airspace
- UAS Weather
- UAS Loading and Performance
- Crew Resource Management
- NOTAMS & TFR's
- Radio Communication
- Emergency Procedures
- Preflight and Maintenance

Once you are ready to take the exam you will need to schedule the test through CATS.

https://catsdoor04.com/cbt/online/login.jsp

From here you will pick the testing location and time that is closest to your location.

You cannot take your phone, watch or other items with you during the test. The testing facility will provide you with everything you will need.

They will provide you the following book.

https://www.faa.gov/training_testing/testing/supplements/media/sport_rec_private_akts.pdf

BEYOND THE BASICS

Get familiar with this book so that you can quickly find what you are needing during the test.

There are 63 questions on the test but only 60 count towards the grading. You will need 70% to pass the exam.

After you pass the test you will need to create an account on the Integrated Airman Certification and Rating Application (IACRA) system.

BEYOND THE BASICS

https://iacra.faa.gov/IACRA/Default.aspx

It may take up to 10 days for your test score to show up on the IACRA system.

You will need to create an application for a Remote Pilot to assign your test score to. Once in the system, the FAA has up to 90 days to process your application. You will receive a temporary pilot certificate until your official one is mailed to you.

The Survey Mapping Made Simple book series contain methods with step by step solutions and examples.

Throughout this book, tips will be given to help explain or give directions on the subject matter.

Internet links are shown throughout this book. These links maybe come broken over time. A search on the internet for the topic at hand will provide new links to the subject material.

Beyond the Basics

Now that you have your Part 107 Remote Pilot Certificate, you are now an expert at survey mapping, right? Wrong! You have only one piece of the puzzle that is needed to become a true professional at using a drone for survey mapping.

You have your Remote Pilot Certification and a drone, so now you are ready to start marketing your survey mapping services. Not really! There are many things that need to be put in order before you start your first project.

I have instructed my students, colleagues and co-workers that using a Drone for survey mapping is much more than just flying it around and taking pictures or videos.

This series of books on Survey Mapping will provide you with the skills and steps to become a true professional at flying and mapping with a drone.

There are six **KEY** elements that are required to be a true professional when it comes to survey mapping.

1. **Beyond the Basics**: FAA Part 107 Rules and Regulations (Book 1)

2. **What's My Mission?**: The right drone for the job! (Book 2)

3. **Time to Fly**: Mission Planning, Ground Control Points (Book 3)

4. **Deliverables**: Orthomosaic, Point Cloud, DSM, Panorama (Book 4)

5. **Emergencies**: In-flight and during active missions. (Book 5)

6. **Operation Standards** (Book 6) Manual, Training and Testing of Remote Pilots.

Part 107 Rules and Regulations

This book covers commercial drone operations related to survey mapping.

If you want information on recreational drone operations, please refer to https://www.faa.gov/uas/getting_started/fly_for_fun/

Here are some highlights of the Part 107 rules as they pertain to commercial operations:

Pilot Requirements:

Must be at least 16 years old
Must pass an initial aeronautical knowledge test at an FAA-approved knowledge testing center
Must be vetted by the Transportation Safety Administration (TSA)

Aircraft Requirements:

Less than 55 lbs.
Must be registered
https://registermyuas.faa.gov/

Operating Rules:

Class G airspace*
Must keep the aircraft in sight (visual line-of-sight)*
Must fly under 400 feet*
Must fly during the day*
Must fly at or below 100 mph*
Must yield right of way to manned aircraft*

Must NOT fly over people*
Must NOT fly from or over a moving vehicle*
* All of these rules are subject to waiver

Here is a more detailed list:

https://www.faa.gov/uas/media/Part_107_Summary.pdf

National Airspace System

The National Airspace System is a busy and complex network of airspace, navigation facilities and airports of the United States.

In the USA, airspace consists of classes A, B, C, D, E (controlled) and G (uncontrolled).

https://en.wikipedia.org/wiki/National_Airspace_System

This book will focus on the airspace that is associated with drones. From the surface to 400 AGL.

A thorough understanding of Aeronautical Sectional Charts is required.

Below is the sectional chart for the Phoenix Metro Area.

BEYOND THE BASICS

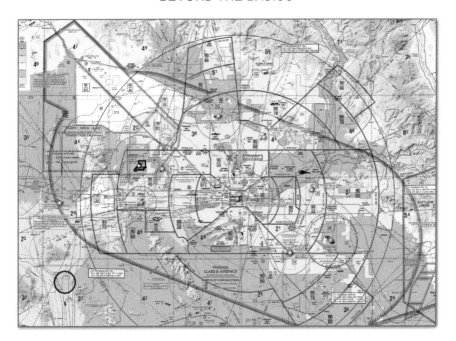

Knowing how to read these charts is a requirement of the Part 107 test as well as mission planning for survey mapping.

The best way to understand what information these charts are showing is to take a cross section through a segment and analyze each airspace classification.

Below is a cross section through the Phoenix Sky Harbor airport.

BEYOND THE BASICS

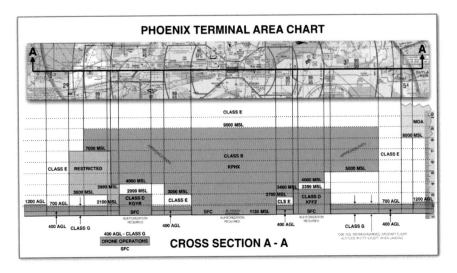

Here is a link to a PDF that can be downloaded and printed as a study guide.

http://www.cc4w.net/remotepilot/
PHX_TAC_Cross_Section.pdf

Drone operations are from the surface to 400' AGL in Class G airspace as shown in the tan color on the cross section.

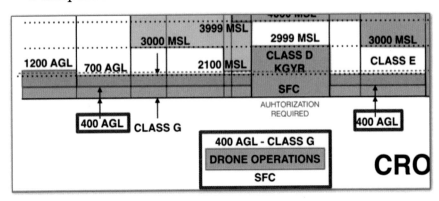

To operate in controlled airspace B, C, D and E, airspace authorization is required which is discussed in detail in the next section.

BEYOND THE BASICS

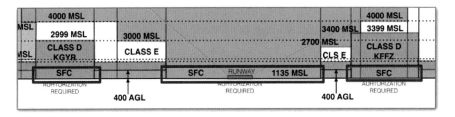

https://skyvector.com/ is a free resource that provides sectional charts throughout the USA. Temporary Flight Restrictions (TFR) are also shown and their time of operation.

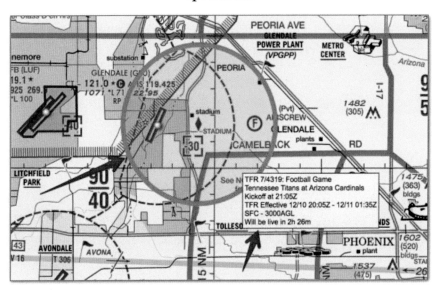

Prior to every flight, Sky Vector should be visited to verify that there are no TFR's near the project site.

Waivers and Airspace Authorization

Airspace authorization and waivers are two separate processes.

Authorization is required to fly in controlled airspace for class B, C, D and E.

BEYOND THE BASICS

Waivers are a regulatory process to waive certain rules such as flying at night, over people, beyond line of site, etc.

The following link will walk you through the process.

https://www.faa.gov/uas/request_waiver/

It is really difficult to get a waiver. The process can be laborious and time consuming. The FAA is approving night time waivers however it takes a lot of documenting for safety, mitigating risk factors, anti-collision lighting and so forth.

Airspace authorization is much easier and most are approved. The 90 waiting period is not conducive for short notice from a client to fly a project.

https://www.faa.gov/uas/request_waiver/request_operate_controlled_airspace/

Each airport facility has developed a grid map that shows the above ground altitude (AGL) that the FAA will approve through the online portal and or LAANC (see below). Each grid is a one minute of arc for both north-south and east-west directions.

Any request for a higher AGL than listed will require a longer review time.

Below is the facility map for Phoenix Sky Harbor International Airport.

These facility maps are dynamic meaning that they will change over time due to airspace requirements. Be sure to check them before each flight.

BEYOND THE BASICS

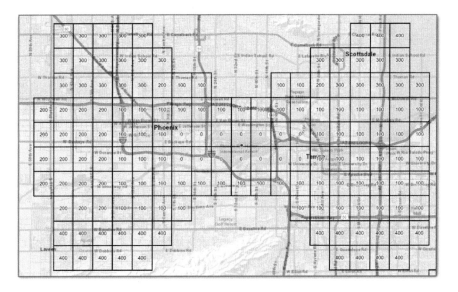

The FAA facility maps can be viewed here:

https://faa.maps.arcgis.com/apps/webappviewer/index.html

The FAA is developing a Low Altitude Authorization and Notification Capability (LAANC) for instant airspace authorization that is being implemented across the United States.

For example: Sky Harbor International Airport is on the LAANC system and will authorize airspace instantly through two approved vendors.

Airmap and Skyward are the two FAA authorized vendors as of the writing of this book. I am sure more vendors will be added over time. All airport facilities will be added to the LAANC system as the infrastructure is completed.

Airmap: https://app.airmap.io/

Skyward: https://skyward.io/

You can read more about the LAANC system here:

https://www.faa.gov/uas/programs_partnerships/uas_data_exchange/

 See the **Sample Mission** at the end of this book for a demonstration of these tools.

Local Drone Laws and Ordinances

The airspace throughout the USA is National Airspace and is controlled by the FAA. States, Cities, Counties, etc., do not control the airspace. Some public agencies think they do but they don't. They can only control take off and landing on property under their control.

Privacy rights, such as flying over private property is controlled by the local laws and ordinances. You need to research and review the local laws to determine when and where you can fly a drone.

For example: Arizona does not have air privacy rights which means you can fly a drone over private property.

Common sense needs to take precedence when flying over any property. If your flight is going to disturb the property owners, then it might be a good idea to not fly over their property. Why create an issue when it can be avoided. It is better to ask permission than to risk a conflict.

NOTAM'S

A Notice to Airmen (NOTAM) is a notice filed with an aviation authority to alert pilots of potential hazards along a flight route or at a location that could affect the safety of the flight.

BEYOND THE BASICS

https://en.wikipedia.org/wiki/NOTAM

A search of the NOTAM for the area that you will be flying in is recommended to see what planned activities are in the area of your project site.

https://notams.aim.faa.gov/notamSearch/nsapp.html#/

Here is a sample search for Sky Harbor Airport.

There are options to view on a map and to show in plain language.

METAR's

Meteorological Terminal Aviation Route (METAR)

METAR is a format for reporting weather information. A METAR weather report is predominantly used by pilots (manned and unmanned) in fulfillment of a part of a pre-flight weather briefing.

Initially the coding and format of a METAR is difficult to understand. With some practice and regular use, they start to make sense.

BEYOND THE BASICS

This link will provide some overall explanation of what a METAR is and what the codes mean:

https://en.wikipedia.org/wiki/METAR

To review a METAR for your project site, use this link:

http://aviationweather.gov/metar

Here is a sample METAR for the Phoenix Sky Harbor (KPHX):

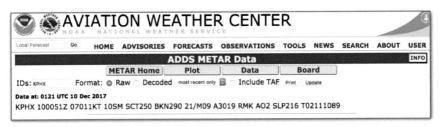

Here is the same sample that has been decoded.

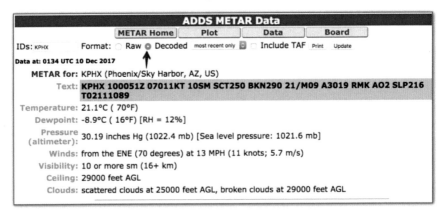

Use the above links to review a METAR for your local airport.

Click the box to Include TAF (Terminal Area Forecast) to view what the forecasted weather will be.

BEYOND THE BASICS

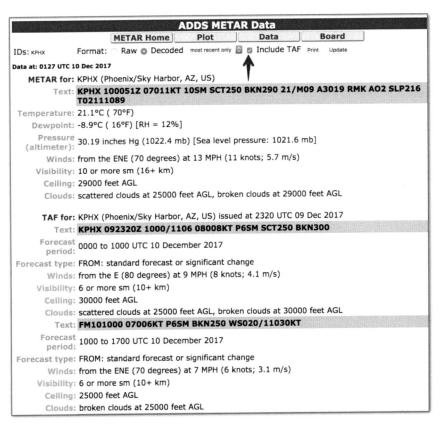

As a private pilot, I review METAR's for the airports that I takeoff from and my destination airport. This gives me vital information such as wind, cloud cover, rain, dew point, etc., that I need to have for a safe flight.

As an unmanned remote pilot, I use the METAR forecast to review upcoming winds for scheduling a flight. The winds, dew point, cloud ceiling heights and visibility are the main items of a METAR for unmanned flights.

 Note: All reports are in Zulu time.

BEYOND THE BASICS

The local weather channel and apps on a mobile device are good resources of information as well.

I have had flights where the nearest airport was calm. When I get to the project site, the winds were over 20 mph so I had to cancel the flight.

I have a wind gauge that I use to verify the wind speed when the weather conditions are questionable.

You need to know what the maximum wind condition is allowed for your drone. Your drone might handle the wind speed, however it will have a huge impact on battery life for the flight. Keep that in mind. If it is windy, align your flight path with the crosswind to avoid draining the battery as much as possible.

BEYOND THE BASICS

 Book 2 "What's My Mission?" will cover the maximum wind speed for several DJI drones.

Another service that the FAA provides for flight planning is

https://www.1800wxbrief.com

UAS flights being operated under a Certificate of Authorization (COA) must file a UAS NOTAM. These operations are known as DROTAM.

Part 107 Remote Pilots are not required to file a UAS NOTAM.

Part 107 Remote Pilots should use the LAANC system for instant airspace authorizations.

Below is a DROTAM that is displayed in purple on https://skyvector.com/

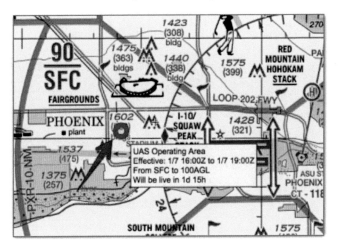

BEYOND THE BASICS

Coordinated Universal Time (UTC)

UTC (*official abbreviation*) is the primary time standard by which the world regulates clocks and time. Check out the link below for more information.

https://en.wikipedia.org/wiki/Coordinated_Universal_Time

The aviation world uses UTC (Zulu Time) because most commercial flights will cross several time zones. It would be way to difficult and confusing to change to local time as each flight cross's a time zone.

To convert local time to UTC use the following formula:

Current local time + time zone offset + 12 if after noon.

6:51 PM + 7 (MST) + 12 (PM) = 25:51 - 24 = 1:51 UTC

The weather forecast sample above was requested at 01:27 UTC. To convert to local time use the following formula:

01:27 + 24 - 12 - 7 = 06:27 PM MST

There are plenty of online apps to make this conversion for you.

Most smart devices will have an option to add UTC in a clock app.

Urban Environment

Survey mapping in an Urban Environment can be real challenging. Most of my urban drone flying has been in the Phoenix Metropolitan area. The map below shows just how complex the airspace is.

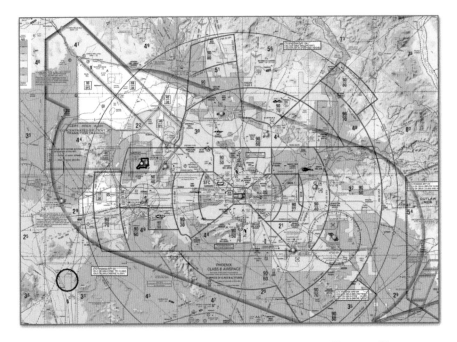

There is Class B, D, G and E airspace as well as Military Operations Area and Restricted Areas to navigate through and around.

There are a lot of manned air traffic in the Phoenix Metro area that include Air Balloons, fixed wing (airplanes), rotary wing (helicopters), vintage airplanes, etc.

Keeping an eye on air traffic while you are flying a drone can be real challenging, near and far away from airports. Manned aircraft are everywhere. I have quite a few hours of flying manned aircraft in this busy traffic area that has given me a perspective that most drone pilots will not experience.

Below is a light day showing air traffic in the Phoenix Metro area.

BEYOND THE BASICS

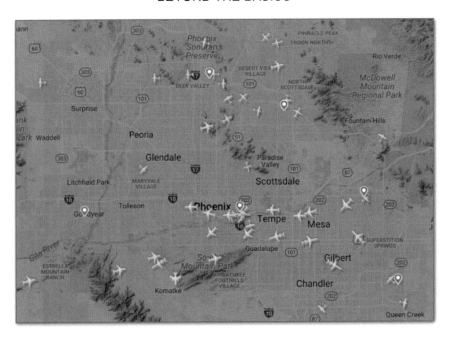

When you are flying in an urban environment, it is very important to listen and scan the air for manned aircraft. I have had to abort several missions in order to lower my drone until the manned aircraft has cleared the area.

What I have experienced is that manned aircraft pilots violate the airspace rules on a daily basis. I have had several fixed wing manned aircraft fly way below the 1000' minimum floor in the urban area. I have had helicopters fly way below the 500' minimum floor in the urban area.

You would think that staying below 400' with a drone that there wouldn't be any issues. This is not the case.

The airspace would be much safer if everyone would just play by the rules. The reality is that is never going to happen.

As a drone pilot, it is your responsibility to watch and listen for other air traffic, manned and unmanned. Manned aircraft have the right-of-way even if they are violating the airspace rules.

The basic rules of keeping the drone in Line of Sight, don't fly over moving vehicles or people still apply no matter where you are flying your survey mapping mission.

Rural Environment

Manned aircraft traffic in a Rural Environment is a little bit lighter depending on the area of the United States you fly in.

Pilots in Training use rural areas to practice in to sharpen their skills and flying techniques.

Manned aircraft floor is at 500' in rural areas. Keep that in mind as you map large areas with a drone.

As a drone pilot, it is your responsibility to watch and listen for other air traffic, manned and unmanned. Manned aircraft have the right-of-way even if they are violating the airspace rules.

The basic rules of keeping the drone in Line of Sight, don't fly over moving vehicles or people still apply no matter where you are flying your survey mapping mission.

Drones and the Public

The Unmanned Aerial Vehicle, commonly known as a drone, is an aircraft without a human pilot aboard.

Drones have been around for a long time. Powered drones have been around since 1916.

Remote controlled model aircraft have been around for a long time as well.

BEYOND THE BASICS

Personal drone manufacturing, both recreational and commercial, started exploding around 2013 due to modern technology.

The public has reacted negatively and positively. More people are using drones for personal and commercial usage.

Several private and government agencies banned the use of drones initially as part of a knee jerk reaction without fully understanding the potential value of this technology.

The FAA has been ordered by congress to incorporate drones into the national airspace system.

Slowly but surely the public and private agencies are coming around to allowing the use of drones for commercial purposes.

At some point in time, the paranoia will pass and drones will be part of everyday life.

I have learned to just tell clients that I will be getting some pictures and videos of the project site and not go into detail as to how I will be taking those pictures and videos. This approach seems to work better than telling them I will be flying around with a drone. They will figure that out when they see it flying around.

Most of my clients are excited to see the technology and the products they receive helps to get them get even more excited.

BEYOND THE BASICS

Flight Logs

The FAA does not require a Remote Pilot to keep a flight log although it is a good idea.

A flight log will document how many hours and what type of drone you have experience with. It will be useful when you are looking for that perfect remote pilot job.

Depending on the drone insurance you have, they may require you to keep a flight log. If you ever have an incident, the insurance company and/or the National Transportation Safety Board may ask to see your flight logs.

I use Microsoft OneNote to keep a digital flight logs. I can update it from any computer or smart device. I keep it current with each flight so I don't forget to log the flight.

Sample Flight Log

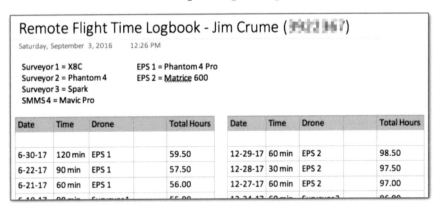

A log book for each drone is also important to keep track of hours on the individual drone, remote pilot who flew the drone and to document maintenance.

BEYOND THE BASICS

Sample Missions
Uncontrolled Airspace

Mission location:

Lat: 33°22'16.65"

Long: 112°11'51.62"

Step 1:

Use Google Earth to review the project location.

Step 2:

Open https://skyvector.com/ and navigate to the above Lat/Long.

Review for any TFR's. (A red circle would indicate a TFR.) The floor of the Class B airspace is at 3000 MSL. Below that is Class E (2999 MSL to 700 AGL) & Class G (700 AGL to SFC) airspace. Cleared for 400 AGL Class G at this location. Note the private airstrip to the Southeast. Probably a crop dusting business. Keep a look out for low flying aircraft.

BEYOND THE BASICS

The upper part of the window will show the Lat/Long of the cross hair at the center of the window. Move the map around with your mouse until the cross hair is at your project location. You will need to convert the Lat/Long to Decimal Minutes.

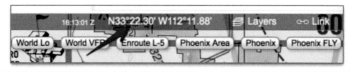

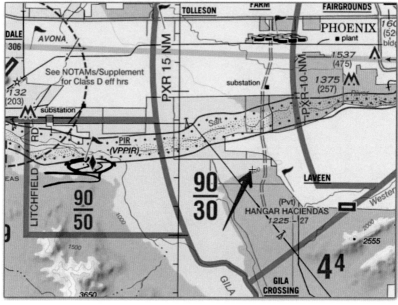

 To learn more about converting Lat/Long to Decimal Minutes, please see "Bearings and Azimuths" - Book 1 of the Surveying Mathematics Made Simple series.
http://www.cc4w.net/ebooks.html

Step 3:

Open https://app.airmap.io/ to confirm Class G airspace.

BEYOND THE BASICS

The mission is a GO. No further airspace authorization is needed.

Controlled Airspace

Mission location:

Lat: 33°23'31.00"

Long: 111°56'29.36"

Step 1:

Use Google Earth to review the project location.

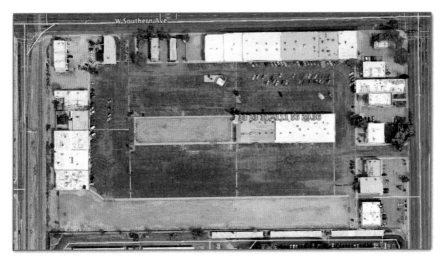

BEYOND THE BASICS

Step 2:

Open https://skyvector.com/ and navigate to the above Lat/Long.

Review for any TFR's. (A red circle would indicate a TFR.) The floor of the Sky Harbor Class B airspace is at the surface. Authorization will be required to fly in this area.

The upper part of the window will show the Lat/Long of the cross hair at the center of the window. Move the map around with your mouse until the cross hair is at your project location. You will need to convert the Lat/Long to Decimal Minutes.

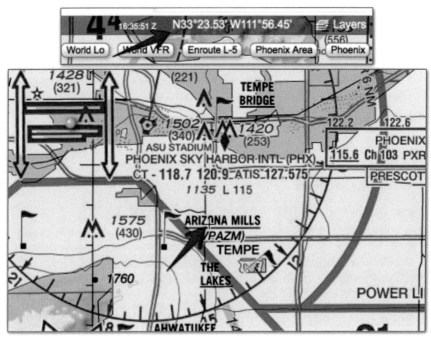

Step 3:

Open https://app.airmap.io/ to confirm Class B airspace.

Click on the map at the project site location. The zone will be selected with relative information shown for that zone.

BEYOND THE BASICS

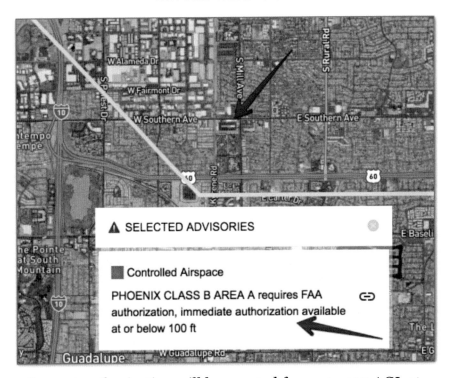

Instant authorization will be granted for up to 100 AGL at this project location. If you need to fly higher than 100 AGL, then you will need to file for an airspace authorization through the FAA online portal and wait for months to get approved.

Once LAANC has been fully implemented, you will be able to apply for higher altitudes and get approval within 30 days.

Book 3 "Time to Fly" will cover the instant authorization process using a smart phone app.

Final Step:

Review the NOTAM and METAR for the nearest airport (Sky Harbor International Airport) to the project site(s).

BEYOND THE BASICS

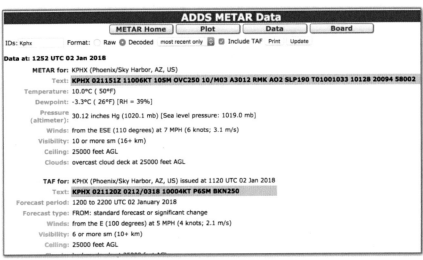

Checklist

As you progress through each book in this series, a check list of items will be added that you will be able to combine for a complete mission checklist. We will cover the checklist in detail in Book 3 "Time to Fly".

	PRE-FLIGHT CHECKLIST
1 - Airspace	
	Review project site in Google Earth
	Review aeronautical chart https://skyvector.com/
	Confirm airspace in https://app.airmap.io
	Authorization needed for Class B, C, D & E Get approval through LAANC or online portal
	Class G - no additional authorization needed
	Review NOTAM and METAR

Conclusion

Knowing and mastering the FAA Rules and Regulations is just the first step to becoming a true professional Remote Pilot for Survey Mapping.

Professional survey mapping is much more than just knowing how to fly a drone.

This series of books will provide a step-by-step process to completing and mastering the use of a drone for survey mapping.

BEYOND THE BASICS

Other books in this series

Digital and **Printed Editions**
Survey Mapping Series Training and Reference Books. Designed and written by Surveyors for Surveyors, Land Surveyors in Training, Engineers, Engineers in Training and aspiring Students.

A **New** Survey Mapping Series of books with helpful hints and easy to follow step by step instructions.

http://www.cc4w.net/drone.html

BEYOND THE BASICS

Other books by this author

MATH-SERIES TRAINING AND REFERENCE BOOKS / APPS

Digital and **Printed Editions**
Math-Series Training and Reference Books.
Designed and written by Surveyors for Surveyors,
Land Surveyors in Training, Engineers,
Engineers in Training and aspiring Students.

A **New** Math-Series of books with useful
formulas, helpful hints and **eas**y to follow step by
step instructions.

http://www.cc4w.net/ebooks.html

ABOUT THE AUTHOR

Jim Crume P.L.S., M.S., CFedS, RP

My land surveying career began many decades ago while attending Albuquerque Technical Vocational Institute in New Mexico and has traversed many states such as Alaska, Arizona, Utah and Wyoming. I am a Professional Land Surveyor in Arizona, Utah and Wyoming. I am an appointed United States Mineral Surveyor and a Bureau of Land Management (BLM) Certified Federal Surveyor. I have many years of computer programming experience related to surveying. I became an FAA Certified Remote Pilot in September 2016. I have been using Drone's for survey mapping since that date.

This ebook is dedicated to the many individuals that have helped shape my career, especially my wife Cindy. She has been my biggest supporter. Without her, I would not be the professional I am today. Thank you very much Cindy.

Other titles by this author:
http://www.cc4w.net/drone.html
http://www.cc4w.net/ebooks.html
Follow us on Facebook
Books available on amazon.com